Fire engines of Europe
by

John Creighton

IAN HENRY PUBLICATIONS
1980

Copyright © 1980, John Creighton

64 Transport series

1 A E C: builders of London's buses
2 Ford panel vans
3 Fire engines of Europe

First published
by Ian Henry Publications, Ltd

ISBN 0 86025 846 7

Made and printed in Great Britain by
Lowe & Brydone Printers Ltd., Thetford, Norfolk
for Ian Henry Publications, Ltd.,
38 Parkstone Avenue, Hornchurch, Essex, RM11 3LW

CONTENTS

40-ton mobile crane, Vienna Fire Brigade

European
fire brigades

THE differences between European fire brigades include a variety of factors such as training, number of incidents, educational qualifications, although this book is concerned with different fire appliances. To illustrate the diversity in numbers of emergencies one can consider a few areas – Helsinki has about 2,000 calls a year, Gibraltar 600, London 97,000, Vienna 21,000 incidents, while differences in manning levels vary too; Holland has 3,800 full-time and 23,000 part-time men, the Swiss city of Berne has 80 full-time and 400 part-time men while Athens has only full-time firemen to run the five fire stations, and in the country of Luxembourg there are 90 full-time and about 7,200 part-time men.

Just as brigades differ with reference to emergencies and manning levels, so too do they vary with types of fire appliances – in addition to water tenders and other vehicles in Geneva, there are 8 Jeeps, while Reykjavik (Iceland) has mainly American Fords with Champion pumps and Helsinki has examples of Volvo, Mercedes and Sisu appliances. Naturally, European countries use a mixture of vehicle makes and the state of Andorra is an example of this with two Magirus-Deutz (German), and Berliet (French), Simon Snorkel (UK) and American Jeep in use.

In this book some of the main European chassis manufacturers will be considered, grouped under country headings, and it should be noticed that many firms export models and that chassis can have different types of bodywork; so for instance a fire engine labelled Ford/HCB-Angus has Ford chassis and HCB-Angus bodywork, and this applies to a variety of other vehicles such as MAN/Bachert, Land Rover/Rosenbauer and Volvo/AB Tollarpkaross

A hydraulic platform, Reykjavik, Iceland, and a Dutch fire boat at Amsterdam

West Germany

Introduction

WEST GERMANY has several internationally famous fire appliance manufacturers and the various regions or "Lands" have their own constitution with each responsible for running its fire brigade. The capital, Bonn, has 300 full-time and about 700 part-time firemen with basic training for full-time men lasting six months, and the city's brigade respond to about 30,000 calls a year.

MAN/Ziegler emergency tender

MAN

MAN manufactures the chassis and cab for fire appliances with bodies supplied by a variety of firms.

Water Tenders

The MAN 11-168 HLF chassis used with Bachert and Ziegler bodywork in the LF 16 water tender is 4,100mm in length and the vehicle is powered by a 6-cylinder water-cooled diesel engine, with a crew of nine, 800-litre water tank and various lengths of ladders. The TLF 16, like the LF 16 is found

MAN/Rosenbauer Foamatic 7000

in many European brigades and has a 2,400-litre water tank and Bachert FP 16/8 pump with a maximum output of 3,200 litres a minute.

MAN vehicle transporter used by Falck in Denmark

MAN 3 axle foam tender on 26.320 chassis

Emergency Tenders and Recovery vehicles

The RW3 emergency tender on a MAN chassis is seen in several European brigades with a water-cooled diesel engine, 15-ton winch, generator and 9-metre telescopic lighting mast plus other rescue equipment. The MAN/Ziegler emergency tender has MAN 11. 168 HA-LF chassis and a comprehensive set of rescue gear, while MAN diesel articulated lorries are used by Danish brigades for vehicle recovery work.

Foam and Air Crash Tenders

The F–TLF 600/1000 has a MAN chassis and its tanks hold 6,000 litres of water with 1,600 litres of foam agent, while the Foamatic 7000 has a MAN forward control chassis and Rosenbauer pump. Metz bodywork is also used with the MAN 26. 320 DFA chassis in a 3-axle foam tender with two roof-mounted monitors.

MAN/Simon Snorkel SS300

Magirus-Deutz LF 16

Magirus-Deutz

Magirus-Deutz fire vehicles are found in almost every European country and this company has been involved in fire-fighting matters since 1864 with its products including portable fire pumps, turntable ladders, water tenders/ladders, airport fire appliances, foam tenders and emergency tenders, the chassis, engines and fire-fighting equipment all being made by one company.

Water Tenders

Magirus-Deutz TLF 16

Magirus-Deutz produced a small pumping vehicle in two versions, the "TSF" (for a crew of five plus driver) and the "TSF (T)" for a crew of two plus driver, and the main function of these vehicles was to set up an effective water relay over a distance of 150 metres from the source of water to the fireground. A more versatile appliance was the pumping appliance with a crew of eight plus driver and these "LF" vehicles include the LF 8 and LF 16, the former having a four-wheel-drive chassis, front-mounted Magirus pump with a delivery of 1,600 litres/minute at eight bars and accommodation for a driver and nine crew.

The LF 16 appliance is found in many European countries, having a crew of nine, with a water capacity of 800 or 1600 litres, and rear-mounted pump capable of deliver-

Magirus-Deutz Tro TLF 16

ing 3,200 litres/minute at 8 bars. The vehicle carries lengths of hose, breathing apparatus sets, suction hoses and various lengths of ladders and is indispensable as a standard rural or city fire appliance, and Frankfurt adapted one of these machines as an emergency tender, adding telescopic floodlights and generators. Magirus-Deutz, however, manufacture the HLF 16, weighing 12 tons, which has normal fire-fighting materials, foam compound and fixed generator set and floodlight mast as standard equipment.

Magirus-Deutz Tro LF 2000

The well proven TLF pump water tenders are longer than the LF series and the TLF 8 is designed for a crew of two and driver, with a water capacity of 2,400 litres, being found in many European brigades. The TLF 8 vehicles are available in 4-wheel drive or road versions with air-cooled Deutz diesel engines and rear-mounted Magirus pump delivering 2,100 litres/minute at 8 bars, while TLF 16 appliances can deliver 3,200 litres a minute at 8 bars. The TLF 16 has room for six men accommodated in a spacious compartment and the water tank is made out of glass fibre reinforced plastic capable of carrying drinking water when required.

Another popular fire appliance produced

Magirus-Duetz DL 18 turntable ladder

Magirus-Deutz DL 30 turntable ladder, Holland

by Magirus-Deutz is the Tro TLF 16 dry powder appliance found in Holland, Germany and many other countries and relying on water, powder and foam as extinguishing agents. The dry extinguishing equipment includes 750kg of powder while the FP 16/8 pump can deliver water at 3,200 litres/minute at 8 bars. Munich Fire Brigade has had over a dozen Tro TLF appliances in service for many years and the dry powder extinguishing equipment makes it a popular choice for industrial brigades in many European countries.

Dry Chemical and Foam Tenders

The Magirus-Deutz 310 chassis is a popular choice for foam tenders and the Tro LF 2000 has 2000 kilograms of dry powder. The ZLF series of fire appliances extinguish fires by foam and the Magirus-Deutz range includes ZLF 24/32, ZLF 24/65, ZLF 40/120; the ZLF 24/32 has 3,200 litres of foam com-

Magirus-Deutz DL 44 metre turntable ladder

10

pound and a maximum speed of around 90 kph while the Tro ZLF 40/100 is a dual-purpose vehicle with 4,000 litres of foam compound, 6,000 litres of water and 1,000 kilograms of dry powder.

Turntable Ladders and Hydraulic Platforms

Although turntable ladders had been known since 1802 (Regnier in Paris) their practical introduction did not come about before 1892 when the first Magirus-Deutz turntable ladder was developed, having an extension of 25 metres. With increasing extension lengths proving difficult for hand-operated ladders, motor-operated systems came into use, and in 1904 Magirus built the first steam-driven ladder for Cologne and the first petrol engine-driven automotive turntable ladder in the world (with mechanical drive of all ladder movements) was supplied to the town of Chemnitz in 1916 by Magirus.

Lift in use on a 44 metre Magirus-Deutz turntable ladder

Magirus-Deutz, with Simon hydraulic boom

Equipment: Magirus-Deutz RW emergency tender

Magirus-Deutz KW 20 mobile crane

Subsequently the firm made rapid advances in turntable ladder constructions, in 1938 producing a 38-metre ladder equipped with the first serviceable set of ladder sections in welded steel construction. In 1950 Magirus built the then highest turntable ladder in the world, the "52+2 metres" ladder, and three years later Magirus were the first in Germany to introduce a turntable ladder where all actions were controlled by hydraulic means and had hydraulic and electric safety devices. Later ladders had rescue cages at the end of the ladder, and today one can even see a Magirus 50-metre turntable ladder, while throughout the world Magirus ladders feature prominently in fire brigades; Melbourne, Australia, has a 44-metre ladder, Bilbao also uses this machine and in 1980 Greater Manchester Fire Brigade purchased two 30-metre turntable ladders.

The DL 18 is a manually operated 18-metre ladder with a crew of three and total weight of 5.2 tons, while the popular DL 30 is found in most European cities, having accommodation for a crew of six and weighing over 13 tons. The DL 30 ladder assembly is hydraulically driven and the positively controlled rescue cage at the top of the ladder can carry two men or 180kg, and levelling devices on this vehicle keep the ladder assembly level at all angles from $-15°$ to $+75°$. The LB 30/5 has three axles, weighs over 21 tons and has a built-in platform while the DL 37 is equipped with a rescue cage and weighs 15 tons, carrying six men.

The DL 44 metre turntable ladder is hydraulically driven and is particularly useful for rescues from tall buildings, with a lift which slides the length of the ladder, proving invaluable when rescuing people since they do not have to walk down the ladder. With a wheelbase of 4,800mm and maximum speed of 82 kpm the DL 44 turntable ladder weighs 17 tons and has a 150-litre fuel tank. The DL 50 turntable ladder has three axles, weighs 22 tons and extends to 50 metres, being found only in a few European brigades such as Vienna and Belgrade

Specialist Vehicles

Magirus-Deutz produce a variety of special vehicles including the emergency tenders RW1, RW2 and RW3, the smallest RW1 ("Rüstwagen 1") weighing 75 tons, having a crew of three and built on an all-wheel drive chassis. Equipment includes a portable generator and winch, found also on the RW2 whose weight is 11.5 tons and emergency gear includes a telescopic floodlight mounted at the rear and an optional 2-ton crane. The larger "Rüstwagen 3" weighs 16 tons and carries heavier cutting and hydraulic gear than the RW1 and RW2 with an engine output of 171kW (232hp).

Among other special vehicles produced by Magirus-Deutz one can consider the SW 2000, a hose-laying appliance while the WAB has interchangeable body units for different incidents such as chemical or pollution jobs, with units carrying the appropriate equipment. A useful appliance when lifting apparatus is needed may be seen in the KW 20 mobile crane whose lifting capacity is 20 tons, and this 3-axle vehicle is found in many full-time and part-time fire brigades in Germany. It is common for some RW emergency tenders to carry an inflatable rubber rescue boat, but some European brigades also maintain the Magirus-Deutz water rescue vehicle (Wasserrettungswagen) weighing 11 tons and carrying a large rescue craft with a crane for lowering it in the water and rear-mounted telescopic floodlighting to assist at night incidents.

Airport Fire Appliances

Many European airports rely on Magirus crash tenders in the event of an accident and a popular model is the SLF 24/40 carrying 3,500 litres of water and 350 litres of foam compound, while the SLF 24/100 has 9,000 litres of water and 1,000 litres of foam compound. The larger FLF 60/130 and FLF 80/200 are found at major airports and this last machine has 18,000 litres of water and 2,000 litres of foam compound, crew of four and four axles.

Mercedes-Benz 508 D chassis used as small pumping vehicle

Mercedes-Benz

Mercedes-Benz offer a wide chassis range of fire-fighting appliances and the company works in close co-operation with internationally famous firms which specialise in fire-appliance bodywork. Mercedes-Benz vehicles are in use in brigades throughout the world.

Mercedes-Benz/Bachert TLF 16

Water Tenders

Small pumping vehicles are found in many areas of Europe and the 508D chassis is used as the base for a light-pumping vehicle while the L8/LAF911 B/36 chassis has *Bachert* bodywork and front-mounted pump, with this LF 8 proving popular as a first-line appliance. The large LF 16 has a Mercedes-Benz fire-truck chassis 1017 F/4×2 and 800-litre water tank with bodywork and pump by Bachert. In the TLF range the TLF 8 has Bachert bodywork, crew of three and a pump delivering 1350 litres a minute at 8 bars, while the TLF 14 has Mercedes-Benz LP 913 chassis, crew of six and 1,400-litre water tank. The larger TLF 16 water tender has Mercedes-Benz

chassis and is frequently seen in European brigades with Bachert bodywork, 2,400-litre water tank, 4×4 all-wheel drive and a pump delivering 3,200 litres of water a minute at 8 bars. The Mercedes-Benz 2632 AF/6×6 all-wheel drive chassis has Bachert bodywork, 5,000-litre water tank, 600-litre foam tank and crew of six, while a TLF 24/50 has 1626 Mercedes-Benz chassis, and Bachert bodywork and pump delivering 3,200 litres a minute at 8 bars.

The Austrian firm of *Rosenbauer* also equip Mercedes-Benz fire appliances and the Mercedes-Benz 1626 chassis has Rosenbauer bodywork and roof monitor which may be raised if necessary, the 1924 chassis also has Rosenbauer parts and roof-mounted monitor. The popular TLF 3000 has Mercedes-Benz 1113 chassis, 2,500-litre water tank and 200-litre foam tank while the

Mercedes-Benzz 1626 chassis & Rosenbauer monitor in low and raised position (right) and (below) Rosenbauer vehicle on M-B 1924 chassis

Mercedes-Benz TLF 3000 on 1113 chassis

Mercedes-Benz foam tender on 2632 chassis

Mercedes-Benz/Vanassche water tender on 1017 chassis

TLF 2000 has the 1017 chassis with Rosen-bauer bodywork and roof-mounted monitor.

The firm of *Ziegler* also supply bodywork for Mercedes-Benz chassis including the TLF 16 with a pump delivering 2,800 litres of water a minute at 8 bars, and the TLF 24/30 appliance with Ziegler bodywork has a 1632 AF/4×4 all-wheel drive chassis, 400 litres of foam and 3,000 litres of water. The TLF 28–35 has Mercedes-Benz 1113 chassis, Ziegler bodywork with tanks holding 3,500 litres of water and 400 litres of foam compound. Roof-mounted monitors are useful in that they allow for impressive distance and velocity, with the TLF 24/50 for instance, having a monitor delivering water/foam at 1,600/2,000 litres a minute, having Ziegler bodywork and Mercedes-Benz 1719 chassis.

Metz bodywork appears on a variety of chassis makes and the TLF 16 range includes a model with Mercedes-Benz 1017 chassis, Metz bodywork, a 3,600mm wheelbase, crew of six and 2,650-litre water tank In the TLF 24/50 range of fire appliances, Metz supply the bodywork for the Mercedes-Benz chassis 1719 and this popular vehicle has a crew of three, 5,000 litres of water, 500 litres of foam compound and the usual fire-fighting gear.

Mercedes-Benz chassis for water tenders

are found in countries outside Germany, and in Belgium the firm of *Vanassche* provide bodywork for a number of water-tender chassis including the 1017 F/4×2 and 1017 F4×4 makes which have 2,000 litres of water, Godiva pump and crew of eight. Examples of this "Autopompe Semi-Lourde" are found in Kortemark, West Flanders, while Mercedes-Benz 1319 F/4×2 chassis with Vanassche bodywork is in several areas of Belgium including Stavelot in Liège province.

Mercedes-Benz 3 axle foam tender on 2624 chassis

Foam and Air Crash Tenders

Mercedes-Benz chassis are frequently used in the construction of foam tenders and air crash tenders, and the 2632 chassis has Rosenbauer equipment and bodywork with a 4,000-litre water tank and 700-litre foam tank. The larger ULF 4500 has two roof-mounted monitors, this machine having a 1419 Mercedes-Benz chassis and 4,500 litres of water plus 500 litres of foam, and Rosenbauer bodywork. The 1926 Mercedes-Benz chassis is used for the ULF 5000/1000 foam tender whose impressive equipment includes 1,000 litres of foam compound, 6,000 litres of water and 1,000 kilograms of dry powder.

Mercedes-Benz Foamatic 4500

The 2632 chassis is the basis of a vehicle with 11,000 litres of water and 1,000 litres of foam compound, while Rosenbauer also equip a 3-axle, two-monitor foam tender on the Mercedes-Benz 2624 chassis. This same bodywork firm provides monitors and bodywork for a variety of Mercedes-Benz air crash tenders such as the Foamatic 4500 and 5000 with crew of six, while the Foamatic 8000 crash tender had Mercedes-Benz 2632 chassis and a Rosenbauer pump delivering 3,000 litres a minute at 10 bars, and the Daimler-Benz diesel engine produces an impressive speed of 96 kph.

Among the Bachert range of foam tenders with Mercedes-Benz chassis, the Sch TLF 5000 has a 1719 chassis, 5,000 litres of foam compound and the roof-mounted monitor can deliver 2,000 litres a minute with a 50-metre range. The S-TLF 6000

Mercedes-Benz 1419 chassis, Metz DL 30 turntable-ladder

Mercedes-Benz/Bachert foam tender discharges a water/foam mixture at 2,400 litres a minute over 60 metres while the S-TLF 6500 has a foam-agent tank with a capacity for 6,500 litres.

Turntable Ladders and Hydraulic Platforms

The Mercedes-Benz 1419 F/4×2 chassis is used in conjunction with a Metz DL 30-metre ladder in the production of a turntable ladder found throughout Europe, having a 4,200mm wheelbase, six-cylinder diesel engine and crew of three. Mercedes-Benz also produce a vehicle with the Simon Snorkel SS 263 whose working height is 27.5 metres, and the 1113 chassis is sometimes seen with a turntable ladder.

Simon Snorkel SS263 on Mercedes-Benz chassis

Turntable ladder on Mercedes 1113 chassis

The 1619/4×2 chassis supports a hydraulic platform supplied by Ruthmann (MTS 210) and the vehicle has a 5,200mm wheelbase, working height of 22.40 metres and crew of three. The large Ruthman hydraulic platform (MTS 300) is mounted on a Mercedes-Benz 2626 K/6×4 chassis and is powered by an eight-cylinder diesel engine having a working height of 32 metres. The Kirsten Telelift is mounted on a special chassis with Mercedes-Benz components and has a 12-cylinder diesel engine, while the rescue platform accommodates nine people.

Emergency Tenders

Mercedes-Benz have a number of emergency tenders carrying equipment for use at incidents such as road or air crashes and oil pollution and the RW2 has *Bachert* bodywork with Mercedes-Benz 1017 AF/4×4 all-wheel drive chassis and six-cylinder engine. With a crew of three this vehicle is equipped with geneator, hydraulic winch, 7-metre lighting mast and a rubber boat on the roof. The RW3 emergency tender is also seen with Bachert bodywork and Mercedes-Benz chassis with six-man crew and similar gear to that on the RW2, with the exception of the winch which is a 15-ton version as opposed to the 5-ton on the RW2.

18

Mercedes-Benz Kirsten Telelift hydraulic platform

Mercedes-Benz Foamatic 8000

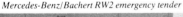

Mercedes-Benz/Bachert RW2 emergency tender

The LAF 1113/4×4 all-wheel drive chassis is used with *Rosenbauer* bodywork in the production of the popular RFC 11 emergency tender found in several countries including Austria and Germany, and powered by a six-cylinder diesel engine this vehicle has modules or containers which are removed from the middle of the appliance by a rear-mounted crane. Each module has equipment necessary for various jobs such as decontamination, chemical or salvage incidents and before leaving the fire station the appropriate module is lifted into the vehicle, which also has a generator winch and 7-metre lighting mast. The Mercedes-Benz 1017 chassis also appears with Rosenbauer bodywork as an emergency tender and equipment includes a crane and rubber rescue craft.

Ziegler supply body parts for Mercedes-Benz emergency tenders with 1017 chassis and the RW oil/crane appliance has equipment designed to combat oil spillages, plus lighting mast, fully hydraulic crane capable of lifting 470 kilograms, with power coming from a Mercedes-Benz diesel engine.

Specialist Vehicles

Mobile recovery cranes are common in most European brigades and within the Mercedes-Benz range the 2626 K chassis is used with a Holmes 750 crane, and night rescue incidents are assisted by a Mercedes-Benz lighting vehicle with Polyma floodlighting unit which has a working height of

12 metres with six halogen floodlights and three tripod searchlights. Large scale emergencies demand an on the spot control unit and parts of Germany have the Mercedes-Benz 0 305 converted single decker bus with usual control and communication facilities, while the Mercedes-Benz 350 SE saloon car is sometimes used as a smaller control unit having also lighting and road traffic accident rescue gear.

Forest Fire Appliances

The TLF 1250 has Mercedes-Benz UNIMOG 125/416 chassis with Rosenbauer bodywork and 1,250 litres of water while the TLF 1300 has a crew of five and 1,300 litres of water. The Mercedes-Benz Unimog 125/4164×4 chassis is also seen in conjunction with Bachert bodywork as an all-wheel drive forest fire-fighting vehicle with pump, water tank and winch. The Unimog TLF 8/18 pump water tender with a 1300 L chassis and Ziegler bodywork is well suited for rough terrain and forest fires, and Mercedes-Benz Unimog together with Metz bodywork have produced an impressive vehicle which has flexible chassis and coil suspension, allowing maximum axle articulation over rough ground. With a crew of five this machine can fight forest fires and is also valuable at road traffic accidents in that it can leave a motorway and travel cross-country to reach the front of a hold-up.

Mercedes-Benz RFC 11 emergency tender with removable container (below) TLF 8/18 forest fire appliance

Mercedes-Benz/Unimog cross country fire fighter

Opel Blitz with Metz bodywork, 1949

Opel

The first Opel fire engine was produced in the early 1900s and the firm's truck chassis is used by several fire appliance manufacturers including Metz, Rosenbauer and Bachert.

Water Tenders

The famous Opel Blitz water tender has been in production for some time and the machines illustrated give examples of this fire vehicle with Metz bodywork in 1949 and 1952 and Bachert bodywork in 1951. In the 1960s, models included Opel's first semi-forward control vehicle, with Rosenbauer equipment, while the 1970s saw German brigades using a conversion of a factory-built van with six-cylinder engine.

Opel Blitz LF8 with Metz bodywork, 1952

Faun

Faun airport appliances are large vehicles based on designs created to deal with crashes of larger aircraft. The Foamatic 11000 air-crash tender has Faun chassis LF 910/42 V 6×6 all-wheel drive, a 10,000-litre water tank and 1,000-litre foam tank, with acceleration 0–80 kph in 40 seconds. The large Faun 8×8 appliance is equipped with 18,000 litres of water, 2,000 litres of foam, 1,200 kilograms of dry powder, and is powered by a Daimler-Benz V 10 diesel engine.

Opel Blitz with Bachert bodywork, 1951

Faun Foamatic 1100

Henschel

This firm uses a number of bodies and pumps such as Rosenbauer and Bachert, and in 1968 merged with Hanamog, and a popular vehicle is a water tender on a Henschel HS 100 4×4 chassis, with 2,000 litres of water and a Rosenbauer pump discharging 1,500 litres a minute.

Hanamong/Henschel vehicle

Volkswagen

This famous firm supplies a number of different vehicles which serve as light fire tenders, personnel carriers or command units, and Vienna, for example, has a crew-carrying vehicle, while some Dutch brigades use Volkswagen vehicles for carrying specialist gear such as extra breathing apparatus sets.

Branbridge (UK) base a light fire vehicle on the Volkswagen double cab pickup with a custom built rear half of the body designed to carry four men, while Rosenbauer (Austria) equip the Volkswagen vehicles as fire appliances and the VW/LT 35 is a popular example of this machine.

Volkswagen personnel vehicle

Steyr TLF 4000 water tender

Austria

Introduction

THERE are many examples of fire appliances in Austrian fire brigades including Mercedes-Benz and Volkswagen, and in the country two internationally famous manufacturers have their factories – Steyr who produce chassis, and Rosenbauer whose pumps, bodywork and other products are referred to in many parts of this book. Vienna has almost two million inhabitants and over 20 fire stations with most vehicles built by Rosenbauer on Steyr chassis, with the exception of turntable ladders. Incidents in Vienna are often dealt with by rapid intervention appliances but every job is also attended by a fire unit consisting of five appliances – water tender, turntable ladder, multipurpose appliance, rescue vehicle and officer's car.

Steyr

This firm provides chassis for a variety of fire appliances seen in vehicles in Austria and Liechtenstein.

Water Tenders

The TLF 2000 Steyr is a popular vehicle on the 590/038/4×2 chassis with Rosenbauer foam/water monitor, while the TLF 4000 with Rosenbauer bodywork and roof-mounted monitor has a 3,800-litre water tank. A multipurpose vehicle designed for Vienna fire brigade has a Steyr 790 chassis, 1,200 litres of water, 600 of foam and 500 kilograms of dry powder. A heavy duty Steyr fire tender in Vienna has three axles, weighs 22 tons and is equipped with 7,000 litres of water, 1,000 litres of foam, and the roof-mounted monitor delivers 1,600 litres a minute at 10 bars.

Turntable Ladders and Hydraulic Platforms

The Steyr 1290 chassis carries a 26-metre hydraulic platform while the 690 chassis is often seen in Switzerland and Austria with a 27-metre turntable ladder.

Steyr 27-metre turntable ladder

Steyr multipurpose vehicle

Foam Tenders and Emergency Tenders

The Steyr/Rosenbauer SLF 1300 foam tender can deliver 2,400 litres a minute over 60 metres while the TLF 7800/1000 is found in many brigades with 8,800-litre tank and Rosenbauer foam/water monitor model RM 16. The Steyr/Rosenbauer RF 16 emergency tender is found in Austria and Liechtenstein with gear including rear-mounted crane and front-mounted winch, while the Steyr 690 chassis is used in a combined firefighting/rescue tender. The 790 chassis is sometimes seen in a hose layer with container by Dlouhy and is popular in many brigades in Europe, including Vienna.

Steyr hose layer

Steyr 3 axle heavy duty water tender

24

Switzerland

SWISS fire brigades are organised at Canton level and manning numbers vary, with Geneva, for instance, employing 200 full-time and 600 part-time firemen to operate over 25 water tenders, six-wheeled escapes and numerous special vehicles. Fire appliance makes in Switzerland include Mercedes-Benz (for example in Zurich), Saurer (in Basel), Jeeps which often tow large wheeled escapes, and Vogt vehicles made in Oberdiessbach, Switzerland.

The Swiss firm of Mowag produce four-wheel drive vehicles using their own chassis or converted Chrysler (Dodge) chassis with the Chrysler V8 engine. A popular water tender has eight men and weighs 4.5 tons, while the Mowag W 500 water tender of the early 1970s has a 1,200-litre water tank and Rosenbauer pump.

Saurer

Saurer produce chassis for use with a number of fire appliances including a water tender with Rosenbauer bodywork and 2,000-litre water tank, air-crash tenders with the 6 GAF – LL chassis, and turntable ladder with Magirus ladder. Examples are found in many brigades, including Zurich, where appliances have the Saurer CT 30 135 DIN/PS engine and several water tenders in Istanbul have a Saurer chassis.

Jeep towing wheeled escape
Sauer/Rosenbauer water tender

Berliet fourgon pompe tonne

France

Introduction

FRANCE has a population of some 55 million and there are 12,000 full-time and 200,000 part-time firemen in the country, with a further 9,000 under the control of the military in Paris and Marseilles. Manning levels and types of appliances vary, and the Nord Département (including towns such as Lille and Dunkirk) has 800 full-time and 4,800 part-time men, while Paris has 7,000 full-time men who respond to about 135,000 incidents a year, using a variety of appliances including about 60 turntable ladders, 15 hose layers and over 160 water tenders.

Berliet

This is one of the main French manufacturers of fire vehicles and in 1971 joined together with Citroën and Guinard to form the "Camiva" company.

Water Tenders

The range of Berliet chassis used with water tenders is vast and examples include chassis 500 KE for a small vehicle, the "fourgon pompe tonne leger" with 1,500-litre water tank, while the 770 KB 6 chassis is seen in a larger water tender. The Berliet "fourgon pompe tonne" is found throughout France including the "Alpes-Maritimes" brigade in the south of the country. Other chassis include the 881 KB which is on a vehicle carrying 5,000 litres of water, 500 litres of foam and crew of three, while the Berliet L 64/8R 4×4 chassis is used for water tenders in semi-rural areas of France with a crew of three and Peugeot engine. Paris is an example of a brigade with the Berliet "fourgon pompe" with 1,000 litres of water, 160 litres of foam, rescue gear and breathing apparatus sets.

Dry Powder and Foam Tenders

The Berliet chassis is used with many foam tenders and the 881 KB chassis has Camiva parts, 3,500 litres of water, 500 litres of foam and Camiva pump delivering 1,000 litres a minute. The GR 280 chassis is the base for a foam tender often seen at French oil refineries, having a crew of three, foam tank and monitor delivering 4,000 litres a minute at 7 bars over an effective distance of 60 metres.

The L64 8R 4×4 is used with a foam tender in industrial areas with petroleum fire risks, having 1,200 litres of water and 1,000 litres of foam compound with a monitor whose capacity is 1,500 litres of premixed foam a minute. Larger oil refinery tenders have the Berliet chassis GBH 260 6×4 and a maximum speed of 86 kph, being driven by a Peugeot engine and carrying 10,000 litres of foam compound. The smaller refinery tender on GR 280 chassis has roof-mounted monitor and a top speed 100 kph, while the 440 K Berliet chassis has 1,500 kilograms of dry chemical powder and crew of three.

Turntable Ladders and Hydraulic Platforms

Berliet supply a chassis for various turn-table ladders found in many parts of Europe and among these is the GAK series commercial chassis of the early 1960s, used in conjunction with a hydraulic platform for heights up to 25 metres. The Paris Fire Brigade has over 60 turntable ladders ranging from 18 to 45 metres and many are equipped with a Berliet chassis, with ladders by Metz, Magirus or Riffaud, and one of the few 45-metre turntable ladders in Europe is based in Paris, with chassis by Berliet.

Berliet GR 280 foam tender

Berliet 30-metre turntable ladder

The popular 25-metre semi-automatic turntable ladder is used by many French brigades, having a Berliet 620 KB chassis and a four-cylinder diesel engine and crew of three. The larger 30-metre automatic turntable ladder is built on a Berliet 770 KB 6 ultra long chassis, powered by a six-cylinder diesel engine and accommodates a crew of nine, while some French brigades have examples of Berliet/Simon Tracma hydraulic platforms.

Berliet 500 KE forest fire appliance

Forest Fire Appliances

Forest fires present a tremendous hazard for fire brigades throughout Europe and the summer of 1979 saw devastating blazes in the south-east of France where fires (some 12 miles in length) caused problems for residents and holidaymakers. During high summer, 1979, fire-fighting reinforcements were brought to the Toulon area from as far away as Lyons and Paris while two of the worst affected Départements were Alpes-Maritimes and Var, containing the Riviera

Heavy forest fire appliance with berliet L 64/8 R 4x4 chassis

Berliet/Simon Tracma hydraulic platform

towns. In Alpes-Maritimes there are 500 full-time and 1,400 part-time men, while Var Département has 300 full-time and 1,400 part-time firemen with each Département having between 14,000 and 18,000 incidents a year.

A popular forest fire machine has the Berliet 500 KE 4×4 chassis with a Renault pump and six-cylinder petrol engine while equipment for this appliance includes a 1,900-litre water tank, 80 metres of hose, rear-mounted hose reel, and an electric winch is offered as an option. A larger forest fire appliance has the Berliet L64/8R 4×4 chassis and a 4,500-litre water tank, three-section ladder and accommodation for five men and can reach speeds of over 80 kilometres an hour.

Specialist Appliances

The Berliet chassis is often found as a base for hose-laying vehicles and Paris Fire Brigade have 15 of these appliances equipped with 2,000 metres of hose joined together so that they can be laid while the machine is in motion. A Berliet chassis 440 K long, with Camiva equipment is used as a rescue tender in several French brigades, being powered by a diesel engine and having a variety of rescue equipment.

Citröen

This famous company produces a number of chassis for use with emergency vehicles in several European countries, and is part of the Camiva Company, whose base is at St. Alban-Leysse, near Chambéry.

Water Tenders

The first attendance ("premier-secours") vehicle is found throughout France and is made by several firms including Citröen-Guinard, Citröen-Maheu, Saviem-Maheu and Unic-Maheu. The first attendance vehicle which is illustrated is a Citröen "premier-secours" with equipment by Maheu-Labrosse of Grenoble, carrying 600 litres of water, 40 litres of foam agent, hose reel (80 metres) and 30 metres of hose. The rear-mounted pump can deliver 1,000 litres a minute at 15 bars and there is a crew of six to man this machine, whose popularity in France is proven by Paris having around 60 of these appliances.

Citroen first attendance vehicle (right)

28

An early Hotchkiss first attendance vehicle

Peugot light ambulance, with Brussels Fire Brigade

Renault two man fire appliance

Turntable Ladders

Citröen provide a chassis for many turntable ladders such as Metz and Riffaud and throughout France there are examples of Citröen chassis used in this way.

Specialist Machines

Among the many specialist vehicles used in European fire brigades which have a Citröen chassis one can consider the C 35 model used by Dutch brigades as a water rescue unit. This appliance has a crew of five who are trained in underwater searches, and equipment includes a rubber rescue craft, wet suits and resuscitation gear.

Hotchkiss

The Hotchkiss chassis is used for a variety of fire appliances in France including the early models of "premier secours" vehicles, while Guinard also equipped Hotchkiss water tenders in France. This same company used a Hotchkiss chassis in the construction of forest fire appliances in France carrying 2,000 litres of water and hose reel.

Peugeot

This firm supplies a number of emergency vehicles to European brigades including ambulances operated by French fire brigades and the brigade in Brussels. The J7 Peugeot is a popular ambulance and is seen throughout France.

Renault

Renault vehicles are often used as staff cars in European brigades and in 1979, for instance, the Brussels fire brigade bought a Renault R4, while several Dutch brigades use the Renault 12 for officer use – "Officierswagen". An interesting small two-man fire appliance manufactured by Renault is found in Paris for use in narrow spaces inaccessible to larger machines, and equipment includes 50 kilograms of powder, 20 metres of 20 mm hose, rescusitator, 3.75-metre ladder and radio.

29

Riffaud 24-metre turntable ladder

Riffaud

The "Société des Échelles Riffaud" has been manufacturing fire ladders since 1858 and today Riffaud ladders range from 10 metres to 31 metres and are found on a number of chassis including Saviem and Berliet. Riffaud are also involved in cross country and forest fire appliances and the company's products are found in many European brigades.

Turntable Ladders

The Riffaud semi-automatic turntable ladder is available in 18, 21 or 24 metres with some models of the 24-metre ladder having a rescue cage, this ladder moving from 0° to 75° in 40 seconds, while it takes 60 seconds to turn 360°. An interesting 24-metre Riffaud ladder is mounted on a Salev chassis and used for cross-country purposes or in spaces inaccessible to larger appliances; Paris Fire Brigade has ten of these versatile machines operated by one man and capable of 35 kilometres an hour, powered by a Perkins diesel engine, with an overall weight of 7.25 tons. The Riffaud 30-metre auto-

Riffaud ladder on Salev chassis and (below) 30-metre turntable ladder

Riffaud CCMF 2000 cross country vehicle

matic turntable ladder is equipped with a cage at the top of the ladder for rescue purposes which can carry weights up to 170 kilograms, and French brigades often use this ladder in conjunction with Citröen, Berliet or Saviem chassis.

Cross-Country Vehicles

The popular Riffaud CCMF 2000 appliance has excellent cross-country ability and a 2,000-litre water tank plus pump and rear-mounted hose reel.

Saviem

Water Tenders and Foam Tenders

Saviem have provided a variety of chassis for European fire appliances, particularly in France, with bodywork by a number of firms such as Drouville and Camiva, and the popular first attendance vehicle in France is often a Saviem-Maheu appliance. Useful for rough terrain, the Camiva/Saviem crash tender has Saviem SPS 4×4 chassis, 1,200-litre water tank, 150 litres of foam com-

Saviem/MAN foam tender

31

Saviem hydraulic platform

pound and 25 kilograms of dry powder, with a maximum speed of 80 kph.

A large foam tender has a Saviem/MAN 19240 HA 4×4 chassis and is equipped with 6,000 litres of water, 720 litres of foam compound, having a top speed of about 95 kph.

Turntable Ladders and Hydraulic Platforms

The Saviem chassis is often used for turntable-ladder appliance and among the dozen semi-automatic 24-metre ladders in Paris, several have a Saviem chassis with Riffaud ladders. The Paris brigade has three hydraulic platforms to assist over 60 turntable ladders and these too have a Saviem chassis and a MAN engine producing speeds in excess of 70 kilometres an hour. These machines weigh 20 tons, have three axles and can rescue people 25 metres from the ground while fire fighting is assisted by a 35mm nozzle in a monitor situated in the rescue cage.

Other vehicles

Many European fire brigades operate an ambulance service, and a Saviem chassis is often used with Camiva, and the SB 2L 35N chassis can carry a bodywork designed to accommodate six stretchers or 12 seated people. The Saviem SP5 4×4 chassis appears in many forest fire appliances, often with a four-cylinder Renault engine and Camiva pump delivering 500 litres a minute at 10 bars. A mobile crane with three axles and rear-facing crane is produced by Saviem in conjunction with Pinguely and used by some French fire brigades where heavy lifting gear is required.

Some other French Manufacturers

Heavy mobile cranes feature in many European fire brigades, being useful at incidents where, for instance, vehicles or railway coaches have to be moved. The firm of Demag supply a crane and chassis for a variety of these vehicles and the "Camion-Grue" number 4 has a Deutz engine, four axles, weighs 32 tons and is usually accompanied by a lorry which carries a variety of slings, chains and hooks. A chassis by Beaufrère carries a rear-facing jib on a vehicle which weighs 26 tons, has a Berliet engine and two winches.

Italy

Introduction

THE companies of Fiat, Lancia and OM united with Unic and Magirus-Deutz to form Iveco, producing a specialised range of trucks and buses, and in January 1980, Iveco (UK) Ltd. was created from Fiat Commercial Vehicles Limited and Magirus-Deutz (Great Britain) Limited.

The Italian Fire Service is organised at Government level on a national basis, Rome having over 20 fire stations whose incidents are approximately 9,600 fires a year and 10,000 special service calls such as road traffic accidents. Other Italian cities have impressive fire cover with Bologna having 350 firemen for 5,600 calls a year, Milan 15,000 emergencies a year, while Venice has 15 fire stations, this city using fire boats in addition to road vehicles.

Fiat

This famous company produces a number of chassis for European fire appliances including foam tenders, turntable ladders and water tenders.

Water Tenders

One of the most popular water tenders in Europe, the TLF 2000 Fiat, has Rosenbauer bodywork and 75 PC Fiat chassis with 4×4 all-wheel drive, 1,700-litre water tank, 200-litre foam tank and a Rosenbauer pump delivering 2,800 litres a minute at 8 bars. This versatile Fiat machine performs well on roads and rough terrain and has a crew of three to operate equipment which includes a roof-mounted foam/water monitor with an output of 1,200 litres a minute, and front-mounted winch. A smaller version, the TLF 1300 is available with 1,300-litre water tank and a crew of six. Fiat water tenders with Antonicelli pump are common in Italy and the Fiat 160 NC/B HP 210 chassis is found in many Italian brigades with the pump delivering 4,500 litres a minute and a 5,000-litre water tank, and the Fiat 100 NC also proves a useful appliance in several Italian systems.

Fiat TLF 2000 water tender

Fiat 100 NC water tender

Turntable Ladders

Fiat chassis are used in conjunction with Magirus-Deutz ladders to produce a popular fire-fighting vehicle and examples of the 30-metre ladder are seen throughout Italy, where Bologna has four machines, Genoa four, Rome seven, while Turin has two of these tall appliances. The 26-metre Magirus/Macchi ladder on Fiat chassis is a common turntable ladder in several Italian brigades.

Foam/Dry Powder Appliances, Emergency Tenders and Cranes

The Fiat chassis is used in conjunction with a number of bodywork manufacturers in the construction of foam appliances for use at airfields or oil refineries and the Fiat 682 N3 4×2 chassis with Bergomi equip-

26-metre Magirus/Macchi turntable ladder on Fiat chassis

Fiat emergency tender

ment is a popular machine. The 130 NC/B chassis carries an Antonicelli pump capable of delivering 8,500 litres a minute and foam compound at 250 litres a minute, while the three-axle Fiat 300 PC/B chassis has 1,000 kilograms of dry powder, a 5,000-litre water tank, 2,000-litre foam tank and Antonicelli pump delivering 200 litres of foam a minute. It is interesting to note that Fiat fire appliances are found outside Italy, and Mayo Fire Brigade (Eire) has a Fiat emergency tender. Many Italian brigades have mobile cranes, and two popular examples are the Fiorenkini and Cristanini, both on Fiat chassis, each capable of lifting 16 tons.

OM

The OM/Chinetti foam tender is found in several Italian oil refineries while the OM 160 water tender is a popular fire appliance in the country, and the model "Tigre–M2S" has a 3,000-litre water tank.

OM 'Tigre-M2S'

DAF FF 1600 DT chassis with Saval-Kronenberg body

Holland

Introduction

THE municipal law and fire service law place the responsibility of fire brigades in the hands of local authorities in Holland and the country has about 3,600 full-time and 22,500 part-time firemen. Two major Dutch firms involved in fire appliance manufacturing are DAF and Alkmaar with the latter specialising in hydraulic platforms.

DAF

This famous company manufactures a number of fire appliance chassis which are used with various body manufacturers

Water Tenders

DAF water tenders can have Berwi bodywork while Saval-Kronenburg equip the DAF FF 1600 DT chassis with bodywork and this water tender has a 3,600mm wheelbase and is often seen with rectangular flashing beacons. The FFV 1800 DT chassis is sometimes seen with den Hartog bodywork while Rosenbauer provide the body for the FF 1600 DT 360 water tender chassis.

DAF/Metz DL 30-metre turntable ladder

36

DAF/Saval-Kronenburg foam tender

Turntable Ladders and Hydraulic Platforms

The DAF FF 2200 DU chassis is used in the construction of hydraulic platforms with bodywork and hydraulic platforms often being supplied by Alkmaar (Philips). A popular turntable ladder found in Holland has DAF FF 2000 DH 485 chassis and Metz DL 30-metre ladder and this vehicle may be powered by a six-cylinder diesel engine.

Emergency Tenders and Foam Tenders

DAF/Berwi 4×2 chassis emergency tenders carry a wide range of rescue gear and the DAF FF 1600 DT chassis is sometimes the base for Bikkers bodywork when used as an emergency tender. The DAF FF 900 BDD chassis serves as a large control unit while the FF 2800 DKS chassis with Saval-Kronenburg bodywork is a popular foam tender with a six-cylinder diesel engine and 4,000mm wheelbase. The FFS 2800 DKT chassis also has Saval-Kronenburg bodywork and serves in parts of Holland as a three-axle foam tender.

Alkmaar

This firm produces a hydraulic platform mounted on a variety of chassis makes including Magirus or Mercedes-Benz, and the Alkmaar Lift PH 22/3 has a maximum working height of 23.5 metres with a weight of 9,000 kilograms (excluding chassis and bodywork). The cage capacity is 350 kilograms and the articulated arm consists of three sections stowed in a compact travelling position, and all movements of booms, rotations and stabilisers are hydraulically operated. Optional extras include searchlights on the working cage, monitor in cage (3,000 litres a minute) and a water curtain round the cage with 20 nozzles (450 litres a minute) for the protection of occupants.

Tatra Universal fire tender

Eastern Europe

Introduction

FIRE BRIGADES in Eastern Europe use a number of fire appliances including Zil in Russia, Tatra, Avia, Skoda and Praga in Czechoslovakia and Zuk in Poland. Yugoslavian brigades use several types of machines including those produced by Tam while in East Germany there are examples of Barkas appliances, many of which are converted vans with portable pumps.

CZECHOSLOVAKIA
Tatra

This chassis is used with a number of body types including Karosa and Rosenbauer and the Tatra Universal Fire Tender has the 148 PPR 14 chassis, diesel engine and 6×6 all-wheel drive, with crew of three and equip-

Tatra SLF 8500

ment including 3,000 kilograms of dry powder, 2,000 litres of water and 2,500 litres of foam compound. The Tatra/Rosenbauer PL6 6000 has three axles and V8 diesel engine while the Tatra SLF 8500 has a roof monitor and two removable ones at the rear of the vehicle. The Tatra four-axle ULF 3000/10000 air-crash tender with Rosenbeau body has a roof-mounted monitor delivering foam over 60 metres (*see right*).

Avia and Skoda

The Avia firm makes chassis used in the manufacturing of a water tender accommodating eight men, and the Skoda 706 RTHP chassis acts as a base for a vehicle with crew of eight, 3,500 litres of water and 2,000 litres of foam. Some Czech brigades also use vehicles produced by Praga and Laurin and Klement, the Praga A150 being a large vehicle with three axles.

POLAND

Zuk

The Zuk fire tender is popular with many Polish brigades with the range of vehicles including small pumping appliances, the Zuk A–14 has a four-cylinder petrol engine, 200 metres of hose and portable pump while the A–15 often tows a trailer containing extra lengths of hose.

RUSSIA

UAZ

There are basically three types of Russian fire appliances with one variety having 3,000 litres of water, another 4,000 litres while some machines are designed to work in extreme hot or cold climates. The UAZ–450 A is a light fire tender with portable pump and a four-wheel drive forward control chassis is used for a water tender with 1,610-litre water tank, 55 litres of foam and 1,200 litres a minute rear-mounted pump, and if necessary water and foam compound may

be mixed before delivery. This vehicle has a heating process which provides warmth for the cab and also pumps and water tank, to combat freezing.

Zil

Zil produce a number of chassis seen in eastern Europe, particularly Russia, including a water tender chassis, and the ZIL 131 6×6 chassis used with a 30-metre turntable ladder.

YUGOSLAVIA AND HUNGARY
Tam, Raba

The Tam 4×4 and 4×2 chassis is popular with several Yugoslavian brigades while the Hungarian company, Raba, produce a variety of fire appliances including some mounted on the Austro-Fiat chassis, while larger machines have the Krupp chassis.

Csepel-Ikarus

Some Hungarian brigades use Csepel-Ikarus fire appliances with the 344 type proving popular as a four-wheel drive cross-country vehicle with a 2,000-litre water tank.

Raba/Rosenbauer fire tender

Scandinavia

DENMARK

Introduction

FIRE cover in Denmark is provided by "Falck" (a private organisation) and also municipal brigades, with Falck giving ambulance and vehicle recovery services in addition to providing fire appliances. There are about 140 Falck stations manned by 4,000 full-time and 1,600 part-time men trained in each of the three emergency services. In Norway there are 1,900 full-time and 1,400 part-time firemen and Oslo has 10 fire stations with most appliances having Mercedes-Benz chassis and Rosenbauer bodywork and pumps.

In Stockholm, Sweden, there are examples of Scania and Volvo fire vehicles and the city has eight fire stations run by over 400 men, while in Finland there are 3,100 full-time and 2,400 part-time men with most appliances being Sisu or Volvo. In Helsinki there are 350 full-time and 800 part-time men to run 23 fire stations and there are various levels of entry to the brigade, depending on educational qualifications, with all entrants receiving similar basic training.

Nielsen

Danish fire brigades use a variety of vehicles such as Toyota hose layers, MAN, Mercedes and Ford, and for instance the large port of Esbjerg has among its vehicles four MAN water tenders, eight Mercedes ambulances, five Toyota breakdown vehicles and several Volvos. The main Danish manufacturer is H. F. Nielsen in Haslev where chassis are obtained from abroad and bodywork is added by the company, and appliances include the 8/8 and 16/8 water tender, and hose layers. The 8/8 vehicle has Mercedes-Benz L 608 D/2950 chassis, front-mounted Ruberg pump and crew of 6, while the Nielsen firm's 16/8 vehicle has MAN 13.168 H chassis, rear-mounted Ruberg pump and 2,600-litre water tank, while the Mercedes-Benz 1019 F chassis is also used.

Nielsen/Mercedes-Benz 8/8 water tender

SWEDEN

Scania

This company produces a variety of chassis for use in the manufacturing of water tenders, turntable ladders, hydraulic platforms and other vehicles.

Scania/Metz turntable ladder

Water Tenders

The L81 S42 chassis with E-Rasch-Olsen bodywork is found in several brigades including Ringerike in Norway and this machine has a 3,000-litre water tank, front-mounted pump and crew of six. The same chassis also appears with Bruk (Oslo) bodywork and Oslo fire brigade has examples of this water tender which has a Rosenbauer pump and 6,000 litres of water, while the Scania LB 81 S 38 chassis often has bodywork by AB Tollarpkaross (Sweden) and the water/rescue tender has fire-fighting and rescue gear. Hebra Brand AB also provide bodywork for Scania LB 81 S chassis and this machine has a front-mounted pump.

Scania/Hebra Brand articulated emergency tender

Turntable Ladders and Hydraulic Platforms

Built in 1979 by Metz, a turntable ladder stationed at Fredrikstad, Norway has Scania L 81 S chassis, diesel engine and Metz DL 30-metre ladder, while an earlier model with LD 81 S chassis is found in the Stockholm Fire Brigade. The Scania LS 86 S chassis and LS 80 S chassis are used for hydraulic platforms with High Ranger and Skylift being popular makes of hydraulic platform, while the Scania L 110 S 54 chassis with Simon Snorkel hydraulic platform is found in Bergen, Norway. The Scania LS 80 S chassis with a 24-metre High-Ranger platform is seen in the town of Västeras, Sweden.

Scania SBAT 111 emergency tender

Other Scania Vehicles

The LBS 86 S chassis sometimes carries a large water tank and front-mounted pump while LB 81 S chassis and Hebra Brand AB bodywork is seen as an articulated unit with Örebro Fire Brigade, as smoke exhauster and emergency tender. The SBAT III chassis forms the base for an all-wheel drive emergency tender with hydraulic crane and telescopic lighting.

Volvo

This internationally famous firm manufactures a variety of vehicles with the first truck built in February 1928 starting a reputation for excellence and quality. Early fire appliances include the LV–81 (1939) with Gothenburg fire brigade, the LV–94 and LV–102 light rescue crane also used by this brigade in the 1940s.

Light Pump Appliances

The Volvo C202 has a 500-litre water tank, Ruberg M5–CB pump and 12 metres of suction hose, plus hose reels, being used as a first-strike appliance, with a maximum speed of 115 kph. The C 303 is suitable for cross-country work with a top speed of 160 kph while the C 306 forest fire fighter is a three-axle machine with bodywork by AB Tollarpkaross, 1,000-litre water tank for water or premixed water/foam. The Ruberg HS pump delivers at 500 litres a minute and with a crew of seven, top speed of 90 kph this versatile appliance is powered by the Volvo B 30 six-cylinder engine.

Volvo C303 fire tender

Volvo C306 forest fire tender

Volvo FB13 chassis, Hebra Brand bodywork

Volvo N720 water tender

Water Tenders

Volvo water tender chassis are used in conjunction with a variety of bodywork manufacturers and Stockholm fire brigade has water tenders on the F613 chassis and F84 chassis with Hebra Brand bodywork, and examples built in the 1970s have front-mounted pumps; Tollarpkaross also supply bodywork for Volvo chassis and Moss fire station, Norway, has an example of the Volvo N 720 chassis with a water tender which has a 3,000-litre water tank, Allison MT 650 automatic transmission and usual fire-fighting and rescue gear.

Volvo chassis has Junsele bodywork and the Solna fire brigade has a water tender on Volvo F 85 chassis, Junsele bodywork, being built in 1971. Volvos are not restricted

to Scandinavia and brigades in Belgium
have water tenders with the Volvo F 613
4×2 chassis and bodywork by the Belgian
firm of Vanassche, with either a Godiva or
Bekaert centrifugal pump.

Air-Crash Tenders

The F 12 Fire/Crash Tender is available in
4×4 or 6×6 versions (both with all-wheel
drive) and the F 12 has a minimum ground
clearance of 310mm, accelerating from 0–80
kph in 35 seconds, powered by a Volvo
direct injected, six-cylinder TD 120 turbo-
charged diesel engine.

FINLAND

Sisu

This firm builds a variety of commercial
vehicles including fire appliances such as the
L-Series with distinctive bonnet and front-
mounted pump. The popular six-cylinder
water tender found in Finland has a Valmet
611 CSA engine and Allison M7–650 automatic transmission and front-mounted
pump, while the Sisu 6×2 hydraulic platform is another popular fire appliance in
Finland.

Sisu water tender, with distinctive bonnet

Sisu water tender, with Valmet 611 CSA engine

United Kingdom and Eire

Introduction

THE provision of fire brigades in England and Wales is the responsibility of administrative counties, and control is exercised by the Home Secretary who ensures that the operational efficiency of brigades is maintained. Approximately one-third of U.K. firemen are employed on a part-time basis and full-time recruits follow a 3-month training course.

Eire has a mixture of full- and part-time men, and in keeping with many European brigades, Eire has different levels of recruitment with most senior ranks requiring a university degree or equivalent, while fire prevention officers usually need architectural or engineering backgrounds and qualifications.

Fire appliance manufacturers in the UK export models to some European brigades and most fire vehicles in UK and Eire brigades are made in England.

AEC

AEC produced several varieties of commercial vehicles including lorries and fire appliances. It later merged with Leyland. AEC fire appliances include pump escapes with Merryweather and HCB-Angus body-

Austin/Miles water tender

work, and the AEC chassis is a common choice for the turntable ladder appliance with Merryweather ladders.

Austin

Austin chassis have been used by different European brigades and bodywork types include Kronenburg (Holland), used with the British Austin loader chassis, and Miles.

Bedford

The Bedford chassis is used in conjunction with a number of bodybuilders, and Bedford appliances appear in fleet lists of many brigades in the UK and elsewhere in Europe.

AEC Mercury/Merryweather turntable ladder, 1967

Bedford/Merryweather

The Bedford KG fire appliance chassis is used in conjunction with Merryweather to form the Merryweather Marquis Eight Type "B" Water Tender, powered by the Bedford 500 series six-cylinder engine, having a 1,800-litre water tank, rear-mounted multi-pressure pump and lockers with rolled shutters.

Bedford/Carmichael (top) and Bedford CFE Type 'B' water tenders

Water Tenders
Bedford/Carmichael

Some Bedford/Carmichael water tenders have diesel engines, crew of six, with the former Southport brigade having examples of this appliance, and the Bedford/Carmichael 1,818-litre water tank fire tender is built on the Bedford EFN3 chassis (petrol engine) or the EPRI (diesel engine), being used in several brigades including the Central Area (Scotland). This vehicle has a six-cylinder diesel engine, two 18.2-metre hose reels and pump discharging 2,273 litres a minute at seven atmospheres.

Bedford/Cheshire Fire Engineering (CFE)

This Bedford KG500 water tender has a six-cylinder engine, 127-litre fuel tank, 1,818-litre water tank and 13.7-metre ladder accommodation. A Godiva UMP.50 multi-pressure rear-mounted pump is fitted as standard with portable pump space in either of the forward lockers. The modern type "B" water tender produced by CFE has a Bedford KG fire chassis and its distinctive looks plus many safety features make it a popular vehicle, and in 1979 North Yorkshire fire brigade took delivery of examples of this machine, having twin square blue repeater lamps at the front and rear in addition to roof-mounted beacons.

Bedford 'Green Goddess

Bedford/HCB-Angus refinery tender

Bedford/HCB-Angus

Among the impressive Bedford water tenders/ladders with HCB-Angus bodywork one can consider the type T 10 on TK chassis, while the K9 type chassis with water tender type "B" is found in many brigades including the Isle of Wight. The modern design of the Bedford KG 500 chassis "Crew Safety Vehicle" has excellent vision for the driver and good safety features. The popularity of Bedford/HCB-Angus vehicles is seen in large orders made by brigades and in 1980 the Strathclyde (Scotland) brigade ordered 20 water ladders with Bedford chassis, HCB-Angus safety cab and the rest of the bodywork by Fulton & Wylie.

Bedford "Green Goddess"

The Bedford Model RLH 4×4 and 4×2 Special Appliance and Fire Tender chassis were used by the Civil Defence and Auxiliary Fire Service, and more recently by the army during the strike by UK firemen. These appliances are usually finished in green and have a Sigmund 4,090-litre a minute pump and often carry a portable Coventry Climax 1,363-litre a minute pump.

46

Bedford/Carmichael vehicle with Magirus turntable ladder

Bedford TK chassis, damage control vehicle

Foam/Air-Crash Tenders

The Bedford TM2600, 6×2 chassis is used as a multi-purpose appliance with facilities for delivering water, foam or dry powder and this Bedford/Merryweather vehicle has impressive equipment including two dry powder discharge guns, two foam discharge hoses and foam branch pipes plus a roof-mounted monitor.

The Bedford/Merryweather oil refinery foam tender has a Bedford KG, 4×2 chassis, Bedford 500 series engine and 4,545-litre foam tank with a manually operated monitor operated from a platform behind the cab. HCB-Angus provide bodywork for a number of Bedford foam and air-crash tenders including the "M" Type chassis for a medium airport crash tender and the "TM" three-axle Bedford chassis with HCB-Angus bodywork is a refinery tender with two roof-mounted monitors.

Other Vehicles

A Magirus turntable ladder is sometimes mounted on a Bedford/Carmichael vehicle with accommodation for a crew of six, while the Bedford "M" type chassis appears with HCB-Angus bodywork as an emergency tender, and Essex fire brigade recently ordered a Bedford control unit with bodywork by Benson.

Glasgow Salvage Corps use the TK chassis as the base for a Damage Control unit built to the specification of the Corps, while Merryweather supply parts for a water tanker, and HCB-Angus also equip a water tanker for use in areas where a static supply of water is not readily available. The German firm of Bachert supply bodywork for the TLF 2000 fire appliance which has a crew of nine.

Bedford/Bachert TLF 2000 fire appliance

British Leyland

Leyland fire appliances made an appearance in the early 1900s with the first vehicle going to Ireland, and during the 1920s the FE series came on the market being followed by the FTI and TLM in the 1930s. Following these were the FK and FT type, and bodywork for Leyland appliances is supplied by a number of companies, while the company also provided a chassis for turntable ladders.

Water Tenders and Pump Escapes

A Leyland/Albion all-steel cab was a feature of the Albion/Carmichael water tender which had a Leyland six-cylinder diesel engine, and the Leyland "Comet" machine was popular in the 1950s with South Eastern Area Fire Brigade (Scotland) one of many brigades possessing this vehicle.

Towards the end of the 1950s the Leyland Firemaster chassis came on the scene with a front-mounted pump, and this pump escape had an engine mounted amidships, and although only a few Firemasters appeared, they included pump escapes, turntable ladders and emergency tenders. Manchester Fire Brigade had examples of Leyland/Carmichael Firemasters with wheeled escapes and front-mounted pump, until, in the early 1960s, production of this model ceased.

Small Pumping Appliances

The Leyland Ladbroke Sherpa 250 pickup is the base for a new small pumping appliance, the "Sherpa Firesprite", useful in confined areas such as shopping precincts and industrial estates. Manufactured by Town and Country Factors (a subsidiary of Ladbroke Group Ltd.) this versatile machine has a 570-litre water tank, pump, hose reel, fog gun and portable extinguishers.

Leyland/Ladbroke Sherpa 'Firesprite'

Simon Snorkels on Leyland chassis MS2400 (above) and MS1600 (below)

Nubian Major crash tender, with Carmichael bodywork

Thornycroft/HCB-Angus Super Nubian Major 1 chassis

Turntable Ladders and Hydraulic Platforms

The Leyland TLM chassis was produced to carry the Metz turntable ladder and had a 14ft 6in wheelbase and six-cylinder petrol engine. A Leyland MS2400 chassis with Cheshire Fire Engineering bodywork and Simon Snorkel SS 300 is a three-axle appliance, and the Leyland MS 1600 chassis is often seen with a Simon Snorkel SS 220 with a maximum working height of 23.5 metres.

Air-Crash Tenders

In 1955 "Scammell" joined the Leyland Motor Corporation and as an important section of the famous commercial vehicle manufacturer, the Scammell company kept its individuality and continued its policy of producing special-purpose vehicles includ-ing its range of air-crash tenders. These include Scammell/Pyrene foam tenders and the Scammell Nubian range of air-crash tenders on the Crash Fire Rescue chassis; the design features of the latest Nubian Crash Fire Rescue chassis reflect 30 years of experience and Carmichael built their first air-crash tender on a Scammell Thornycroft Nubian chassis.

The Thornycroft Nubian Major six-wheel drive chassis was designed as an aircraft crash tender and Thornycroft use a number of bodybuilders such as HCB-Angus, Carmichael, Merryweather, Hestair Dennis and Gloster-Saro, and the Thornycroft/HCB-Angus crash tender has the Super Nubian Major 1 chassis. The recent Scammell Nubian range of appliances include the Nubian 2, Super Nubian, Nubian Major 2 and Super Major.

50

ERF Type 'B' standard water tender

Water Tenders

The ERF chassis with bodywork by HCB-Angus is found in numerous brigades and this Type "B" Standard Water Tender has the "F" series chassis; the dual-purpose tender with ERF "F" series chassis, Model 84CF can accommodate a wheeled escape or 13.7 or 10.7-metre ladders, and London Fire Brigade has examples of this wheeled escape appliance.

Turntable Ladders and Hydraulic Platforms

The "S" series chassis is found with hydraulic platform and turntable ladder appliances with a Cummins V8·555 diesel

ERF

Lorries and fire appliances feature among the range of ERF products with fire vehicles built at Winsford, Cheshire, and in 1977 this division was renamed Cheshire Fire Engineering Ltd. (CFE), remaining part of the ERF group. The ERF chassis is found in many types of fire appliances including water tenders, hydraulic platforms and turntable ladders.

ERF/Simon Snorkel SS hydraulic platform, with Perkins engine

engine and 5.18-metre wheelbase, and some brigades have the SS 85 hydraulic platform with Perkins V8·510 engine. The SS 50 hydraulic platform on "F" series chassis has a maximum working height of 15.24 metres. The ERF "S" chassis is often seen with a Metz turntable ladder (30 metre or 37 metre) having a Cummins V8·555 diesel engine and overall length of 9,449mm. The West Midlands fire brigade has a unique ERF/Anglo/Magirus turntable ladder while London Fire Brigade has examples of the ERF/Metz DL 30 hydraulic turntable ladder.

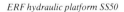

ERF hydraulic platform SS50

FORD

Introduction

The chassis of the Ford Model T was used as a chassis for fire appliances and in the mid-1930s the V8 engine was seen in many vehicles, and today a number of Ford vehicles and chassis are used in the construction of fire appliances. The Transit range is used in the UK and Europe while the "D" series chassis appears in a variety of vehicles with bodywork supplied by a number of companies. Ford produce the D1114 chassis for specialist conversions with a Ford 6.0-litre turbo-charged engine and the D1612 with a Perkins 8.8-litre V8 diesel engine.

Rosenbauer/Ford small pumping appliance

Small Pumping Appliances

The Ford Transit is used as a pumping vehicle in several European brigades, notably in Austria and Germany, and the vehicle is available with a 3.0-litre V6 petrol engine when extra power is required. The TLF 1000 Ford/Rosenbauer is a common pumping appliance with a crew of six and pump discharging 1,100 litres a minute at 8

An early model of Ford/Carmichael water tender/ladder

Ford D1114 chassis, Carmichael bodywork

bars, while the TSF vehicle with Bachert equipment is seen in parts of Germany. Ziegler too equip Ford Transits for European brigades and among UK firms, Cheshire Fire Engineering adapt the Transit 175 as a light pumping vehicle with a 409-litre water tank and Godiva rear-mounted pump.

The Ford "A" series chassis is used in a lightweight appliance in the Cumbria Fire Brigade with a 3-litre V6 petrol engine and is equipped with gear similar to that found on larger water tenders, while the same chassis is utilised by Bridge Coach Works (London) whose multi-purpose appliance has a water tank and Godiva Mark 6 pump.

Water Tenders

The Ford company does not produce any special bodywork for use with water tenders but uses bodies supplied by various manufacturers.

Ford/Carmichael

This popular water tender/ladder is seen in many brigades such as Hereford and Worcester Fire Brigade, with the D1114 chassis, 6.0-litre engine, tilt cab and 1,818-litre water tank; the light alloy dual-stage centrifugal pump can deliver 2,273 litres a minute at seven atmospheres and equipment includes portable pump, ladders and the usual fire-fighting gear.

Ford/Cheshire Fire Engineering

Ford D1114 water tenders/ladders with CFE bodywork have a 6.0-litre Ford engine and transmission with an 1,818-litre water tank and a choice of 13.7-metre or 10.7-metre ladder, plus a 2.28 and 4.58-metre roof ladder. The Ford D1617 water tender/ladder with CFE bodywork has a Perkins V8·540 engine, 1,818-litre tank and measures 7,036mm in length. Equipment is similar to that on the D1114 model and includes a pump and two 54.9-metre hose reels. The D1317 Ford chassis with CFE bodywork also obtains power from a Perkins engine (8.8-litre diesel) and this vehicle has good locker space and easy access to equipment.

Ford/HCB-Angus

The Ford D1000 chassis is used with HCB-Angus bodywork in the production of water ladders, and the Ford D1114 chassis

also appears in the HCB-Angus type "B" water tender.

Ford Water Tenders with European bodywork

Several European body builders use a Ford chassis and at one time Rosenbauer utilised the G398 TA four-wheel drive chassis to manufacture a vehicle with 1,500-litre water tank; Bachert make use of the Ford chassis in the production of the TLF 16 water tender found in several Continental brigades, and there are examples of water tenders in Finland with Ford chassis and front-mounted pump.

Turntable Ladders and Hydraulic Platforms

Among the many vehicles with Ford chassis, the Simon Monitor hydraulic platform has a Ford D1617 chassis, and in 1979 the Royal Berkshire Fire Brigade acquired a turntable ladder based on a Ford chassis.

Emergency Tenders

The range of Ford emergency tenders in Europe is vast and includes many based on the Ford Transit "A" series, and South Glamorgan brigade (Wales) has a Transit/Hoskins vehicle. Cheshire Fire Engineering build an impressive emergency tender on a Ford "A" chassis with pull-out rack from the rear locker and a 3-litre V6 petrol engine. The Ford "D" series is adapted by the firm of Benson as an emergency tender or breathing apparatus control unit, and examples can be seen in Dublin, Staffordshire and West Midlands brigades.

In some European countries the Transit is equipped by continental firms such as Bachert and Rosenbauer, and the Ford Transit/Bachert GW in Germany and GW-01 have rescue gear and decontamination facilities.

Ford D1000 chassis, with HCB-Angus bodywork
Ford/Bachert TLF 16 water tender
Ford water tender in Finland
Ford 'A' series/CFE emergency tender

Foam

The Ford D1617 chassis with HCB-Angus

Ford 'A' series/CFE emergency tender – rear pull-out pack

Ford Granada officer's car

bodywork is a popular oil refinery tender with foam compound and roof-mounted monitor, while towards the end of 1979 the Ford AO609 chassis was used in the production of a smaller oil refinery tender; with a scissor platform carrying a Chubb Slimjet foam monitor this "Firecracker 37" aerial platform has a crew of two, and working height of 11.3 metres.

Other Vehicles

The Ford Transit is sometimes used as a personnel carrier and Rosenbauer equip a vehicle for use in Austria with room for eight men, lengths of hose and portable pump, and some Dutch brigades have Transits for personnel carriers. Bachert adapt a Transit for use as a hose carrier SW 1000, with 800 metres of flaked hose and many brigades use Ford saloon cars for officers' staff cars, with Brussels, for instance, having Ford Granadas.

Specialist vehicles include a Ford/ Willowbrook control unit based on motor coach bodywork with Ford R1014 chassis, seen in London Fire Brigade, while London Salvage Corps has examples of Ford damage control units.

Ford D1617/HCB-Angus refinery tender

Ford Transit personnel carrier

Ford damage control unit

56

Hestair Dennis

The Dennis brothers, John and Raymond, of Surrey, had been bicycle makers in the late 19th century, later moving to the manufacture of vehicles, selling one of their first pumps to Bradford Fire Brigade, and in the early 1900s Dennis Brothers became the main firm supplying fire vehicles to London Fire Brigade, while the Dennis F 24 appliance was the main type in Liverpool Fire Brigade from about 1960. In the late 1970s the firm became known as Hestair Dennis, and fire appliances include water ladders, foam tenders, turntable ladders, control units and emergency tenders.

Water Tenders

Among the range of water tenders, Dennis Brothers Ltd. produced the F12, F15, F24 and F26 machines, with this last model normally having an eight-cylinder Rolls-Royce engine, and the F34 had a 6.5-litre eight-cylinder petrol engine with the

Dennis number 2 pump. The F44 is found in many brigades and the F109 and F108 are often seen with Perkins V8·510 diesel engines.

During the late 1970s many UK brigades, such as Cheshire, acquired the Dennis "R" series with 3.6-metre wheelbase, Allison automatic gearbox and Dennis number 2 pump delivering 2,272 litres a minute, with other pumps offered as an option. The successor to the "R" series has an all-steel cab and this "RS" series has timber or metal body framework of modular construction, with most vehicles equipped with Girling Skidcheck which eliminates uncontrolled skids. During 1979 Greater Manchester Fire Brigade acquired about 30 of these modern machines while in 1979/80 Merseyside, South Yorkshire and West Yorkshire also bought this model.

Hydraulic Platforms and Turntable Ladders

The Dennis F37 and F107 turntable chassis first appeared in the early 1960s and often carried a Metz ladder, while the

Dennis 'R' series water ladder

Dennis 'RS' series water ladder

Other Vehicles

Simon Snorkel SS50, SS70 and SS85 hydraulic platforms are common in many brigades. The Dennis–Snorkel chassis is designed to keep the centre of gravity as low as possible, and power-assisted steering plus an efficient turning circle makes this appliance an asset to any brigade.

The Hestair Dennis/Merryweather foam tender has a Perkins V8·540 eight-cylinder diesel engine and roof-mounted monitor, while London Salvage Corps has examples of Dennis damage control units. London Fire Brigade has examples of the Dennis/ Merryweather emergency tender with a Perkins V8·540 eight-cylinder engine and the usual lifting, cutting and resuscitation equipment.

Dennis damage control unit

Land Rover, Range Rover

European fire brigades adapt the Land Rover to suit their varying needs such as personnel carriers, light pumping appliances or emergency tenders.

Land Rover/Carmichael

The popular "Redwing" series has various models with FT1 and FT2 being similar, each type having hose reels and rear-mounted pump, while the FT3 and FT4 have 340-litre water tanks, lengths of hose and room for five men – three in the cab, two in the rear.

The FT5, however, is fully enclosed, has a crew of five and the vehicle is usually painted red with all stucco aluminium left in natural finish. The FT6 is built on the forward control version of the Land Rover 109 chassis, and has an all aluminium body with fibre-glass roof and accommodation for a crew of five, being particularly suitable for rough terrain; equipment includes a 454-litre water tank, pump, hose reel, portable pumps, lengths of hose, and sometimes breathing apparatus sets.

Land Rover/Branbridge

The Branbridge Mark 1, 2 and 3 conversions are based on the long wheelbase Land Rover chassis with added heavy duty rear suspension. The Land Rover/Branbridge Mark 1 has a crew of three, 456 litres of water, pump and lengths of hose, while the Mark 2 has three extra crew members; the Mark 3 has less covered crew space than the first two types but is equipped with either a pump delivering 1,590 litres of water a minute, or with 272.4 kilograms of dry powder. In spaces with restricted access such as multi-storey car parks and shopping precincts, the low profile Land Rovers are particularly useful as first strike appliances.

Land Rover/Carmichael FT6
Land Rover/Branbridge Mark 2
Land Rover/HCB-Angus light fire appliance

Land Rover/HCB-Angus

The light fire appliance range produced by HCB-Angus includes Land Rovers with a dry chemical unit on the Land Rover 109 chassis with a four-cylinder petrol engine, plus a general-purpose pumper and light pump personnel carrier, and the SPM 400

Low profile Land Rover

light fire appliance. One of the latest Land Rover/HCB-Angus vehicles is the light fire appliance (STD450) with rear-mounted pump, ladders, hose reel and lengths of delivery hose.

Land Rover/Rosenbauer

Among the interesting examples of this vehicle, one with a Land Rover 109 pick-up chassis is seen in Germany and Austria with a four-cylinder petrol engine, front-mounted pump (1,000 litres a minute), 400-litre water tank, three men sitting in the cab, and four more in the rear of the appliance. Another Land Rover/Rosenbauer vehicle is fully enclosed, and mounted on the Land Rover 109 chassis, has a crew of six and 20 lengths of delivery hose.

Range Rover

The four-wheel drive Range Rover estate car came on the market in 1970 and proved popular as a rapid intervention fire appliance, and the Rover/Carmichael Commando is a high performance 6×4 vehicle with good acceleration and cross-country performance, seen in several brigades such as Cheshire, Guernsey and Malta.

The Standard Rover/Carmichael Commando has lockers at the side and rear and can be used as a rescue unit carrying lifting and cutting gear; the water tender Rover/Carmichael Commando has a front-mounted centrifugal pump and water tank plus reels of hose, while the air-crash rescue tender has good acceleration and roof-mounted monitor for delivering foam.

Land Rover/Rosenbauer

Shelvoke and Drewry

Shelvoke and Drewry Limited produce chassis for an impressive variety of Special-Purpose Vehicles (SPV) including water ladders, emergency tenders, hydraulic platforms and turntable ladders and air-crash vehicles.

Water Tenders

The distinctive lines of the Type "B" water tender with Carmichael bodywork are seen in the SPV "Firefighter" vehicle, while a Shelvoke and Drewry 11.5-ton chassis has bodywork by Cheshire Fire Engineering

(CFE). London Fire Brigade are among several who have the WX Series wheeled escape with Cheshire Fire Engineering bodywork.

Turntable Ladders and Hydraulic Platforms

The Shelvoke and Drewry Simon WT Series 6×4 has the impressive Amoco/Simon Snorkel hydraulic platform in use at oil refineries, and the WY 4×2 16-ton GVW chassis (turntable) has Benson bodywork.

Foam Tenders, Emergency Tenders

The West Midlands Fire Brigade have an example of the WX 4×2 emergency tender with equipment including telescopic lighting mast, and the "CX" series 4×4 includes a crash tender with Merryweather bodywork while the Chubb crash tender features in the "CR" 4×4 series.

Shelvoke & Drewry/WY 4x2 chassis (turntable)

Stonefield

Among the Stonefield range a P5000 6×4 was built for Grampian Regional Council Fire Authority fitted with Rexroth winch and mechanical water pump. The Stonefield 4×4 and 6×4 on/off highway trucks with payload capacities of 1.5 to 3.0 tonnes are ideal for specialised needs of fire brigades and the Stonefield/Branbridge rapid intervention vehicle has 456 litres of water/foam, roof-mounted monitor and Godiva pump.

Dodge K850 chassis, with Fulton & Wylie bodywork

Dodge 500 series, K850

Dodge 100 series G13 water ladder

Talbot Dodge

Talbot Dodge (formerly Dodge, Chrysler UK) produce a number of chassis seen in UK and European brigades, including Eire, Belgium, Switzerland and Austria.

Water Tenders

The Dodge K850 chassis has a variety of bodywork suppliers with the Dodge/Unipower/HCB-Angus having accommodation for six men while Strathclyde Fire Brigade (Scotland) has examples of this chassis with Fulton and Wylie bodywork. Bachert also supply bodywork for a Dodge K850 found in Braine L'Alleud, Belgium. In Austria and Switzerland there are water tenders with the Dodge W 300 chassis and Rosenbauer bodywork, while in the UK the Dodge/Commer/CFE G12 water tender has a 1,818-litre water tank and Perkins 5.80-litre engine.

The impressive Dodge 100 series G13 water ladder has Carmichael bodywork, Perkins V8 540 engine and 1,818-litre water tank, while the Dodge 50 series water tender range accepts a wide variety of bodies such as Reeve Burgess six-seater crew cabs and water tender body by Pilcher-Greene Ltd.

Hydraulic Platforms and Turntable Ladders

The Dodge chassis is used in the manufacturing of turntable ladders and a popular UK machine is a Carmichael Magirus 30-metre hydraulic turntable ladder based on the Dodge 16-ton chassis/cab. This vehicle has a Perkins V8 diesel engine and ladder heights can vary, with the range extending to 50 metres, and in the late 1970s Avon Fire Brigade took delivery of two Dodge/Carmichael/Magirus turntable ladders, with the "Hi-line" cab. Dodge chassis, and HCB-Angus bodywork is also seen as a pump hydraulic platform with Simon Snorkel supplying the platform.

Dodge/Carmichael/Magirus turntable ladder

Emergency Tenders and Rapid Intervention Vehicles

The Dodge/Rosenbauer "Cheetah" has 1,100 litres of premixed foam, 250 kilograms of powder and a maximum speed of over 120 kph, making it particularly useful as a first strike air-crash vehicle. A Talbot Dodge road rescue unit with 7.2-litre V8 petrol engine is used by brigades such as Fife (Scotland), and the Brussels brigade has examples of an emergency tender on a Dodge RG 11 chassis with Rosenbauer bodywork, while Ostend uses a high-speed rescue tender (Chrysler B 300) capable of

Dodge/Rosenbauer 'Cheetah'

Dodge high speed rescue tender, Ostend

speeds around 150 kph.

The "Firestreak" rapid intervention foam vehicle on the Dodge Power Waggon W 400 chassis with HCB-Angus body has a maximum speed of over 112 kph and is equipped with 1,000 litres of water/foam premix which can be converted to 8,000 litres of aerated foam when discharged through the monitor.

Specialist Vehicles

Brussels Fire Brigade has several specialist appliances with Dodge components, including a lighting unit and a flood and river rescue vehicle. South Glamorgan fire brigade (Wales) has a Dodge Commando "Rolonoff" appliance at Cardiff together with a Dodge K 850/HCB-Angus foam tender acquired in 1979, while Avon Fire Brigade have a Dodge decontamination/salvage tender.

(above) Dodge flood & river rescue vehicle, Brussels
(below) Dodge HCB/Angus 'Firestreak'